Cow Stories

by James G. Welch

as told to his daughter,

Anne S. Welch

ONION RIVER PRESS

Burlington, Vermont

Stories recorded, transcibed, and edited by Anne S. Welch
Book design by Anne S. Welch
Photo credits: Jonathan Ellwanger, Montgomery Ellwanger, Anne S. Welch, Betty Welch
Special thanks to Margaret Welch for digitizing the slide photos

Onion River Press
191 Bank Street
Burlington, VT 05401

Publisher's Cataloging-in-Publication data

Names: Welch, James G., author. | Welch, Anne S., contributor.
Title: Cow stories / by James G. Welch, as told to his daughter, Anne S. Welch.
Description: Burlington, VT: Onion River Press, 2021.
Identifiers: LCCN: 2021910898 | ISBN: 978-1-949066-67-8
Subjects: LCSH Welch, James G. | Highland cattle--Vermont. | Cattle breeds. | Human-animal relationships. | BISAC BIOGRAPHY & AUTOBIOGRAPHY / Personal Memoirs | TECHNOLOGY & ENGINEERING / Agriculture / Animal Husbandry | NATURE / Animals / General
Classification: LCC SF198 .W45 2021 | DDC 636.2--dc23

Introduction

The Highland Cattle breed was evolved in the difficult and expansive environment of the highlands of Scotland. They were required to be hardy and resourceful to survive. To many people, cows seem dull and slow of wit. The reasons for this are several. First, dairy cattle are visible and common, and the available creatures from which to form this idea. Dairy cattle have been selected for high milk production due to the persistent economic pressure for efficient production. They have also been selected for ability to comply with the demands of a trouble-free milking routine, at least two times a day. There is little room for free thinking or independent action in a milking parlor. Second, most people do not have the opportunity to intensively interact with a group of cattle carrying out their complete life cycle. If one watches people walking on sidewalks in a crowded city, one can also get the idea that humans are also herding animals without much independent action.

As a Professor of Animal Science at the University of Vermont, my primary responsibilities were of course teaching and in addition, doing research with dairy cattle. One of our areas of research centered around the rumination function of cattle. During some of our experiments, it became evident that there were differences between breeds of dairy cattle in their ability to ruminate (re-chew their food). The differences were small, so I went looking for cattle that would be very different from dairy cattle to see if we could find larger differences. I borrowed some Highlands and brought them into our research lab and began a lifelong association with these fascinating beasts.

—James G. Welch

Aster

Aster was the first calf born on the farm. She was born in April, in a snowstorm. When I went to check on the mother cow, I saw from her appearance that she had indeed calved. I knew the calf would be tucked away in a hidey place somewhere, but I had to find it because there was snow on the ground. I looked and looked and looked and finally, I found it. There were a whole bunch of red raspberry bushes up at the top of the house pasture, and the calf was tucked away in those red raspberries. It had a little snow on its back, and it was very difficult to see. I walked up to it. One of your major curiosities is if it is a heifer or a bull, and I wanted to see. I touched the calf, and she let out a bawl of panic, of distress. Her mother was probably 100 feet away and most of the path between the mother and calf was these raspberries bushes, a very thick path. That cow made a beeline for the calf, and me, on a dead run because the calf had given the signal that said, I'm in trouble! I saw this coming, and I did a hasty exit stage left. I moved quickly. Three months later you could still see where that cow went through the red raspberry bushes—her charge.

So, Little Aster was born in the snow. She went on to live to be 19 years old. Because of her status as an old cow, she was the herd leader, the Queen Mother. For most of her 19-year reign, any time I wanted to move the herd, I would just have to go find her and say, "Aster, let's find some new grass." She would pull in behind me and follow me, and the rest of the herd would follow her. Amazing. Just amazing. I never totally understood how she kept her lead status. It was not maintained by physical dominance. Toward the end, she got pretty old and creaky, but she was still the leader. I trained Aster to follow the grain pail, but for a number of years during her life I didn't need it. I would just go find her, and she would swing in right behind me.

Cow Lick

The original two heifers were Nutmeg and Nettles. You remember, one of them made the path through the raspberries to the bawling calf? Well, they had a problem with dogs. One day I heard a dog barking. I looked down in the pasture, and here were Nutmeg and Nettles with the two calves between them. They were standing tail to tail, in a maximum defensive position. That is, one facing one way and the other facing the other way, with the calves between them. They were right out in the middle of the pasture, so this dog had surprised them. I proceeded to scurry to the threat site. I took my cow stick and waved it at the dog and chased it away. After the dog was gone, one of those cows came up and licked me. This was the only time I ever had any—well, call it what you will—contact like that with a cow, where they appeared to appreciate what I did and approached me. Normally, cows like their distance. They don't voluntarily come up to you. The cow generally moves away from you, not toward you, unless you've made a pet out of them. But this one cow came up and gave me a big, abrasive "schwarpfff" on my shoulder with her tongue. A cow's tongue is very rough. It's built for sweeping in grass, and it is very abrasive.

Heifer with Wire on Foot

This story involves a heifer—she weighed 600 or 700 pounds—who got a wire wrapped around one of her hind feet. The loop was such that it was cutting into her skin. When I found her, it was an angry compression wound. The problem is, how do you get this off? The standard behavior pattern of an animal like that is they kick at anything that touches their hind feet. You can't just say, "Stand still!" So, I put her in the chute and went and got a pair of wire cutters. I wasn't sure what I would do next. One of the potentials is I could put a rope loop around the foot and raise it in the air and immobilize her, but that's stressful for her and likely she would thrash around and hurt herself. As I stood there thinking about my options, all of a sudden she just raised that foot in the air so I could reach it. She understood the problem, and she understood that I had the solution for her. She was totally cooperative. I cut the wire off, and she walked away with considerably less limp and healed up very well.

Charlie

You remember the first two cows, the one that made the path through the raspberries to rescue its calf? Well, they needed a bull. Only, they didn't know it yet. (I've got to be careful with anthropomorphism.) I went to a Highland farm, the same one where I borrowed the heifers for the nutrition lab experiments. Have we talked about that? The water bowls? OK, we'll just leave that hanging loose then. It was in the Northeast Kingdom in Vermont, a hill farm with huge rocks and balsam trees in the pasture. Your grandmother and I went up there and looked at bulls. We picked one, and we named him Prince Charles. You know, "Charlie." As in, "Rise and follow Charlie." He was a yearling, probably weighed 650 or 700 pounds. When we brought him to the farm to introduce him, the two cows and their calves were on the top of the hill in the house pasture. The two cows came down the slope and charged poor Charlie. They knocked him down and rolled him down the hill, literally, rolled him down the hill. Cows in general are not welcoming to strangers. That's a basic cow behavior pattern.

Well, Charlie grew and he persisted in his assignment. He eventually bred the two cows and became an outstanding herd bull. There's a whole bunch of things he did. I don't know which to put first. One of his chosen jobs was to be the peacemaker, would you believe, in the herd. That is, cows fight, normally, on a regular basis. They push and shove each other. This is how they work out the dominance within the herd. How that relates to the story about Aster, I don't know. How a lead cow maintains her lead cow position without the physical dominance of the herd, I don't know that. But, that's what happens. There's a lead cow who isn't necessarily the physically dominant animal and then there's a hierarchy of fighting dominance within the cows in the herd. Anyhow, Charlie didn't like cows fighting. Any time cows started pushing and shoving, he would get between them. He basically said, if anybody fights around here they'll have to fight with me. And the cows would have to quit. Some other things Charlie did. In any herd, there will be some cows that like to hang out with the bull. You know, special friends. One time, Charlie was separated from the rest of the herd by three fences. It was during a period we didn't want a cow bred because it would make for calving too early in the spring (late winter is not a good time to calve outdoors in Vermont). Well, Charlie's special girlfriend came in heat. And lo and behold, by the time I got on the scene, he had already broken through two fences. He and this special friend— it might have been Daphne but I don't remember—were going back and forth, not quite nose to nose, on either side of the fence. Charlie was on one side of the fence and the special cow on the other. It was a good fence. It had both barbed wire and electric, and it was being quite effective as far as keeping them apart. I was watching this process. You understand, it would have been nice to put a rope on him and lead him back to where he belonged, but you don't mess with a bull in that condition. So, back and forth and back and forth they went. After maybe fifteen minutes, Charlie steps back and looks, and then he goes up to the gate post that has the

electric fence attached to it. Now, horns do not conduct electricity, so cows do not get a shock through their horns. Charlie hooked his horn behind the gate post and backed up. He pulled out the gate post and then kept on backing up. He pulled out six regular fence posts, and the whole fence went flop on the ground. Then he got to be with his lady love. He walked through where the fence was down and bred her. Just like that. Clever bull. No other bulls ever figured that out. The next year, I put him further away. This story clearly is not a replacement for "Winnie the Pooh."

Another thing Charlie did. Calving season is both wonderful and stressful.

When everything goes right, it is wonderful to watch as this little calf gets dropped on the ground, thrashes around, and gets out of the membrane. The mother licks it and it picks up its head. If everything goes right, within a couple of hours it's standing up and drinking milk out of the udder, sucking away. When that works right, it's just marvelous. Absolutely marvelous. During the time we had cows on the farm, we had over 600 calvings. The excitement and the wonder of that process never left me. Every time I saw it happen—lots of times I wouldn't see it all, but lots of times I would—it was just amazing. Just wonderful, if it all works. But sometimes it doesn't work. This is one time when it didn't. I knew this cow was due to calve. I called the herd, which is how you check on cows. You call the herd to a little bit of grain, and if one doesn't show you know something is wrong. Anyhow, this cow didn't show up, so I went back in the alders to look for her. And lo and behold, I found her. She had just delivered the calf breech, and the calf had died during the calving process. The cow was doing her best to lick the calf off and trying to get it to breathe. The way the cord had been, the calf had been choked off. It was blue, but the mum cow didn't know that was indicative of being dead and she kept licking it. Then Charlie came along. The cow had just about given up, but Charlie moved right in and licked the calf, at first slowly, then faster and faster and faster.

Then he gently turned it over with his huge front foot, and licked the other side faster and faster and faster. Finally, he stood back, gave a huge sigh, put his head in the air, and trumpeted. You could hear the echo coming back across the valley. That's what happened. You can call it what you will, but that's what happened. He just bugled and bugled and bugled. After they lost interest in it, I took the calf away and buried it.

There was another place where Charlie exerted his dominant role in the herd. He never fought with Aster. He never challenged the leadership in the herd as far as the herd going places. But when there was danger, in this case some dogs, he would come to the fore. We had a problem with three dogs coming and worrying the small herd we had at that time. I don't remember how many cows there were, but a few cows, a few calves, and Charlie. I would come across an open spot in the alders where the cows and the calves stood. In the opening, I could see where the dogs had pranced back and forth in front of Charlie. So Charlie was the protector and combatant to keep the dogs away from the cows and calves.

There was evidence of this in the snow a number of times. I tried to scare the dogs away, but I was never successful. As you know, I wasn't there all of the time. The leader was a very big German Shepherd dog. Very wily, very savvy. The minute I showed up or even opened the door to the house, he would start running. One day, I was working the cows in the corral. There were some loggers staying in one of the trailers across the bridge, and they stopped and said, "You haven't been bothered by dogs lately, have ya'?" I said, "Yes, that problem has gone away." They said, "Well, you won't. We took care of them." If I had known who the owner was, I would have called them and told them to keep their dogs at home because they were creating a lot of havoc.

The Beginning

We got an indication in the dairy herd during our research at UVM that there was a genetic difference between breeds as far as the rumination efficiency. But it wasn't a big difference. One day I'm sitting in my office thinking, what could I get that is the most different from dairy cattle? I thought, well, Scottish Highland cattle are more different than anything. Let's compare Scottish Highland with dairy cattle. So, I got five dairy heifers from the UVM farm, and I borrowed five Highland heifers of the same weight. The place where we borrowed the heifers was the same farm Charlie came from. This was pre-Charlie, though. We asked the owner if we could borrow them. He thought, rightly so, that having five of his animals on display at the experimental farm up at UVM would be good, so he said yes. Now, these five heifers came off the same pastures as Charlie. Harsh gravel, rocks and balsam trees, brush, and no barn. They had never been inside a barn. The owner brought them and he led them into the barn. They'd never had a halter on them before, but he led them into the barn. He had just an incredible way with cattle, although I never could figure out

specifically what he did. I'll tell you more about him another time. Anyhow, he brought the Highlands into the nutrition lab, and we put them in tie stalls. The first thing we had to do was teach them how to drink out of a water bowl, which they had never done. They had never drunk out of anything but a brook. The way you teach them is, you let the cow get thirsty and then you press the paddle in the water bowl with your hand. The water runs in, and the cow will drink from the bowl, which is level with her head. Then you press the paddle again, and she drinks again. When the water gets down so the bowl is almost empty, the cow will normally get frustrated and push down with their head. Then that pushes on the paddle, and the water runs in to the bowl. Well, we'd gone through this filling of the bowl and having the Highlands drink several times. They stood back and watched as we pushed the paddle. The bowl filled up and they drank. We pushed the paddle again. The bowl filled up and they drank again. Then we stepped back and said, "OK, you do it." Instead of getting frustrated and pushing the paddle with their head—the expected response—the Highlands picked up their front leg and pressed the paddle with their foot. We ended up having to build a special crib around the water bowl to keep their dirty feet out of it.

After they were getting used to the cleaning and feeding routine, we had to weigh them. They were in tie stalls. And in tie stalls, they are a problem because they kick. They kick anything that goes behind them. This is dangerous, to put it mildly, but it's just their nature. Not just any cow will kick. This is characteristic of a Highland cow. Occasionally, a dairy cow will kick, but generally a dairy cow that kicks goes on the truck. You can't try to milk a cow twice a day that kicks. It's too dangerous. You'd get hurt, and when you're hurt, you can't farm. It would put your whole operation in jeopardy. But the Highlands, they kick. They kick at a hoe, at a rake, at a shovel. For a grazing animal of a primitive sort, that's survival. If there's something behind them that's going to bite, they kick it. But, we had to

weigh them. We had these five dairy animals and five Highlands all roughly the same size, but we needed a body weight on them. So, I look at my lab technician, Ray, and he looks at me. The heifers were in the experimental barn, which is a face-in barn. That is, the animals face in so that when they come out of the stall, they have to back up with their butt to the wall and then go wherever they are going. We put the scale in the middle of the barn. They had to come out of their stall, go around the end of the row of stalls, down the middle, get in the scale, and then complete the circuit by going around the other end of the barn and back to their stall. We decided to start with the dairy animals. None of them were halter broken, so we put a halter on one dairy heifer, backed her out of the stall, and pushed her. One of us was pulling in front, the other pushing behind. One at a time, we got all five dairy animals out of their stalls, around the ends, down the middle, and into the scale with a lot of pushing and tugging. Then we got them back to their stalls and tied up, again, with a lot of pushing and tugging. Well, I look at Ray and he looks at me. We both know we are not going to push the Highlands, because that's looking for a broken leg. I said, "Let's see what happens. Maybe we can kind of herd them around." We untied the neck chain of the first heifer, and it backed out of the stall. Without any pushing or tugging, it went around the end, down the middle, and got in

the scale. The scale had a wooden bottom in it, which was a problem because it went boom, boom, boom, when they stepped on it. But she got in. There was a gate at the back of the box on the scale, and she looked at us as if to say, aren't you going to shut that like you did with the others? So, we shut it and weighed her. Then we opened the front gate, and she walked back to her stall. I looked at Ray and he looked at me and we said, "Wow, look at that!" Four out of the five Highland heifers did exactly that. With the fifth one, the only hitch was when she was going around the end of the stall, she stepped on the big metal plate that's over the gutter cleaner. It shifted and made a noise—went clank, clank—which frightened her, and she

stopped. She just stopped. She didn't know what to do. She didn't want to go forward or back. So, I just touched her with a broom. I didn't put a rope on her or anything, just touched her lightly with a broom, and she went ahead and completed the circuit. All five of them did that circuit after having watched it five times, with no coercion from us whatsoever. I said to myself, those are indeed unusual cows. They'd never been made pets of. They'd still kick, they'd break your leg if you moved in back of them without protection. They were not all kind and gentle. But they were quite willing to adapt to their environment.

Nursing Calf

We've talked about the problem of getting calves started, perhaps ad nauseum. But clearly it is a major problem in the business. This story isn't a usual situation, as most of the calves take right off and things go very well. But the stories tend to settle around problems with the business, which is kind of too bad, but that's the way it is.

This particular cow had a calf that did not get started. I did the usual under those circumstances, which is, I caught the cow, put her in the shoot, caught the calf, held it up next to the cow, and proceeded with the proceedings. That is, I got the calf to nurse and then turned them loose. Usually, that's sufficient for the calf to figure out, oh, that's where dinner comes from. Well, this calf didn't. The next time I checked on them, they were in the house pasture. The cow clearly saw me from halfway across the pasture. She had her calf with her, and it was bonding well, trotting right along with her. But the mother—this is the fascinating part—the mother came and stood right in front of me. Right smack in front of me. Maybe three feet away. I looked at

the calf, and I saw it hadn't nursed since the day before. Basically the cow was telling me, put that kid on again. He hasn't got it. So I talked to the calf, and the calf was very compliant. Standing right there in the middle of the pasture, I pushed the calf while the cow stood still. I worked the udder, made sure the teats were functional, not plugged or anything, and put the calf on. We went through that one more time, but the calf was not very bright. It took three sessions to get it going—one time in the shoot with restraint, and then twice more standing there in the middle of the pasture with no restraint whatsoever. The cow stood still and I pushed the calf up, which is totally unusual. That is, most cows won't let you come near them in the pasture.

Not that they are frightened, just they move away, that's their nature. But that cow came up and stood right in front of me, and I put the calf on while the cow stood there with no restraint whatsoever. Normally, the calves took right off. I don't know what was wrong with this calf. The cow was a successful, experienced mother, but she had been unable to talk this calf into doing it right. Commonly you will see a cow help a calf find a teat. They'll position their legs so the teat is sticking out very evidently. And they'll move their body so the calf is oriented properly and whatnot. An experienced cow will do that. But none of that apparently worked for this calf. It took somebody grabbing it and pushing it to the right place.

Too Warpy

We were having a major problem with stream bank erosion on Tyler Branch. That's always been a problem. A common method of solving this problem is to put rocks—rip rap—on the side of the bank. To do this, I needed a stone boat. You don't buy stone boats, you make them yourself. So I set out to make one. The first thing I did was go to the junkyard and got a piece of a road grader that had a curved structure. You need a piece of heavy steel on the front end to take a chain that is going to pull this very heavy load, and it needs to be able to ride up over other stones and bumps and lumps. It can't be wood, it's got to be steel. So I got that, and then I looked around for what I was going to build the boat part out of, the wooden part. I cut down an elm tree, part of which went into the table that's in my kitchen, that same elm tree. I asked around, where do you get a tree sawed up, and I got a name. I borrowed a pickup truck—I didn't have one back then—and loaded up these two pieces of large elm log. They were heavy, probably weighed half a ton a piece. Then I went off to find this guy. I found the place but when I got there, he was nowhere around.

I looked around and looked around and finally, I saw some people down in the field picking potatoes. I walked down there, and there were two adult men and one teenage boy picking up potatoes. I came up to them and they didn't stop what they were doing at all. I told them the name of the man I was looking for, and both of the adults stood up and each one pointed to the other one. Then they went back to picking potatoes. I'm standing there thinking, what the heck is this? They never said anything, just went back to picking potatoes. I stood there on one foot and then the other, and it became quite apparent that they were not going to stop picking potatoes. So, I bent down and started picking potatoes. This went on for an extended

period, and I'm thinking, this is fine, I'm picking potatoes. They were not talking. We were just working away, picking potatoes. Then all of a sudden, one of them stood up—turned out to be the one that really was the guy I was looking for—pulled out a great big pocket watch, looked at it and said, "Dinner time." It was just noon. All three of them stood up and started walking back toward the house, which was probably a quarter mile away. So I walked back with them. We'd gone maybe 100 yards and the man said, "What do you want?" And I said, "I've got some elm logs that I'd like cut for making a stone boat." He said, "You didn't unload them, did you?" I said no. He said, "Good thing." We walked along a good ways further and he said, "How are you going to build it?" So

I told him about the steel head I had for it and what I wanted from cutting out the elm log. Then I said, "And the rest of it, I thought maybe I'd make a little table or something." "Nah," he said, "don't do that. Elm's too warpy." If you look at our kitchen table, you'll see that he was right. As far as the definition of fine furniture, that kitchen table is sturdy, it has a certain rustic beauty about it, but it's warpy. It's not straight or flat. Anyhow, we went over how I was going to build the stone boat and he said, "What are you going to use for rails?" I said that I didn't think I wanted any rails. "No," he said, "you gotta have rails on it. I'll cut them. I'll make them two inches." We went up and unloaded the logs. He looked at them and scouted how he could cut them and all that, and I went on my way. I went back the next weekend and we went over again how I was going to build this thing. He had some corrections—I won't say suggestions, he wasn't that type—as far as how to put this thing together. I loaded my two-inch elm boards and went on my merry way. I was using the stone boat to haul stones out of some of the stone walls there on the farm and take them down to the bank on the branch where it was eroding badly. One day, I was right beside the road and who drives up but this same man in his pickup truck. He gets out and sees I'm using the stone boat. He says, "You put the rails on?" I said, "Yes, I did." I didn't dare not to. He walked around the stone boat twice and said, "That'll do." Then he got back in his truck and drove away.

Grandmother and Teenager

This story starts with Clover, a big blonde Highland cow. This is way back at the beginning when we had just a few heads. Charlie was the herd bull and was fully grown by this time. He probably weighed about a ton and had a big, heavy set of horns. One morning I go down and check the herd, a small herd at this time, and whoops, there's a half-grown Holstein bull in with them. Only, the Holstein bull is totally beat up. He's all mud where he's been rolled and has huge rake marks where horns had raked his side. His wounds were swelling, all red and raw. He really got hammered by Charlie. He had intentions that Charlie didn't like, and Charlie really beat him up. He was smaller than Charlie, and I felt fortunate that Charlie hadn't killed him. When I opened the gate, the Holstein bull just trotted home. He had had enough. I thought to myself, it's a good thing Charlie was the dominant bull in this case because I really do not want a crossbred Highland-Holstein calf. Well, I didn't think anything more about it until the next spring when, lo and behold, I walk around a bush and there, lying beside Clover, is a bony looking black calf. Whoa, I said to myself, that wasn't

supposed to happen. But it did. Now, there are several side stories to this. Some time ago when DNA testing first came on the scene, somebody in the UVM Med School had a research project looking at the projection of birth defects from DNA analysis. They collected a large group of parents of children with birth defects, and they were going to do great things with DNA testing. But it turned out about 10% of the babies were not as described on their birth certificates. I liken that to what happened to Clover and the small, but effective, Holstein bull. What goes on in the dark of the night behind the bushes is not necessarily recorded accurately. Anyhow, Clover raised this calf successfully. It prospered. At one point, your sister Wynn asked, "Can I have that calf?" In a moment of weakness, I said, "Sure, Wynn, that's your calf." It was a heifer. It grew and became of breeding age and had a calf. It did a good job raising its calf. In fact, it was incredible. For the first part of the lactation, it had way too much milk. The calf would nurse the front teats, and the milk from the back teats would just pour on to the ground. It was the Holstein genetics that made that much milk. Holsteins give five times more milk than Highlands. They have a totally different genetic package. That was a good bull as far as milk production, which is why our neighbor had it, and its daughter produced copious amounts of milk. These two black cows, mother and calf, were at the bottom of the social order in the herd. They routinely got pushed around and shoved away. I'm not sure why. It wasn't just color because there are black Highland cattle in other herds. Anyhow, this second-generation black calf was getting more milk than it knew what to do with, and it grew at an incredible rate. It gave birth to a calf at 18 months. Normally with a Highland cow, you might catch her to calf as a two year old, more commonly at three years old. They are slow. They are conservative because of the environment they are adapted to. Well, the calf with the Holstein genes grew like topsy and got pregnant way too soon, but she calved successfully and everything was fine. The calf got colostrum and all that. Now, during the first few days of a calf's life, the cow will take the new calf off to a secluded area and the calf will find a dark hidey place and lie down and go to sleep. The common saying among people who watch this process is the cow hides the calf. Well, it's half and half. The cow will take the calf to a likely prospect, and the calf will look around until it finds a hidey place, a protected place in the shadow or behind a big rock, that is also protected from the weather.

In a cow herd with calves, evening is a magical time. Just before sunset, the calves usually wake up and go find their mum for the last big nursing of the day. If you are looking for a calf and haven't found it, that's when you look around. One thing that occurs, particularly with young cows, is the cow will go find the calf. She'll go near where she thinks it is and moo a very special moo and say, calf, come nurse. Then the calf will come nurse and maybe play As an aside, evening is also a magical time with the young calves because they all gather. They come out from wherever they are and nurse, and the older ones will play together. They'll run around. Two of them will play "race you to the bush" and run as hard as they can. Then whoops, need to find mom, and they will run back and find mom. Fascinating time when it's occurring.

Anyhow, if you move the herd and a calf is left behind, you often won't find out until the magical time. The cow will go to the gate and bellow. You open the gate, and the cow will go find the calf and bring it back to the herd. That's standard procedure. Well, a couple of days after the teenage mom's calf was born, I had to move the whole herd to a different pasture. But the teenage mom never showed up at the gate to bellow for her calf. I went down and found her and tried to herd her up to the gate, but she didn't want anything to do with it. She didn't know the game at all. It was getting dark, so I went back to the first pasture and looked and looked and finally found the little black calf

sound asleep. It was so full of milk that it wasn't much hungry. I picked up the calf and carried it about a quarter of a mile. It probably weighed 80 pounds, and it was a struggle. I took it to the pasture where the herd was, put it down, and called the cow. Well, the rest of the herd came on a dead run and gathered around me in a ring. I have this black calf between my feet and the cows are circled around us, all looking at the calf, sniffing at it, curious because I'd been carrying it. I'm getting kind of concerned because I'm not sure what's going to happen next. Then, one of the purebred cows touches the calf and that scares it, so it blats. The teenage mother heard her calf blat, and she recognized it and she ran up to the herd. She stood there with a ring of Highland cow butts facing her, and her calf in the middle of the ring. The calf blatted again, and the mother charged through the herd. It just amazed me. Being at the bottom of the social order, this was an intimidating situation for her, but her mothering instinct overcame her fear of being put down. The cows parted to let her through and then re-formed the ring. So now, here's the mother and the baby in the middle of the ring. The other cows weren't being ugly, but they weren't being nice. The mother and calf didn't like the looks of the situation, and neither did I. Then all of a sudden the grandmother, who had been in the back of the ring, charged through the circle. She helped the teenager and calf make their way out of the ring, and then the three of them went off by themselves.

Butchering the First Bull

The first seventeen calves we had sired by Charlie were heifers. On a probability basis, that's less than one in a million, as far as that happening. But, it happened and probably the explanation is there was some glitch in his Y chromosomes, so bull calves didn't prosper in utero. But calf number eighteen was a bull. We named him Edibull. You can spell that any way you want, but the intention is quite clear. Edibull went through his first winter and through his second summer and he grew nicely. In the fall he was all set to go. But then I was talking to our neighbor, and he needed a bull to breed his dairy heifers. I had had hopes of building a market for Highland bulls as sires for first calf heifers because they have very small calves. I had hoped this would be an outlet for my bull calves. So, I talked to the neighbor and he allowed as he would like to try it. I said, "Why don't you borrow Edibull for a while and then we'll bring him back and butcher him." As it turned out, that was a bad idea for several reasons. Number one, as far as furthering the market for Highland bulls, he wasn't big enough to serve the big Holstein heifers. He wasn't quite tall enough, so he didn't do a good

job. But he tried very hard. He was over there for probably two months, struggling bull-fully—not manfully, but bull-fully—to breed those cows and getting nothing to eat but grass hay. I brought him home, and then I thought, OK, Jim, you've got to butcher him. The rest of the herd was clear over in the corral pasture. I shot him and hung him up in a tree by the house. I skinned him out, and I left him to cool overnight. Well, Highlands in particular—cows in general, but Highlands in particular—are very clannish. They also get very upset with blood around. I was sleeping up there overnight and during the early evening, the herd broke through two fences and got up right next to where he was hanging in the tree. And then, they had a wake. It went on all night. It was awful, absolutely awful. They groaned and moaned and carried on all night. Finally in the morning when it got light, they figured out he was dead and gone, and they went back to where their feed was. But it was trauma. I still think about it and I feel awful, just terrible about the whole thing. I packed him off to the freezer locker, and they cut him up and froze him. Do you remember eating him? I later learned that if you want to make an animal tough, that's exactly what you do. You put a young bull to breeding and put him on short rations, which is going to stop his growth. He was like shoe leather. Among the toughest meat I have every tried to eat. That stuff was darn near inedible.

Goat

One of the hazards of having a farm is that sometimes people who have an animal they no longer want—in this case, a goat—will pack it up and leave it off at the farm during the night. Well, somebody left a goat off at the UVM research barns. We thought, OK, here's a goat. I like goats and Ray liked goats, or at least he was willing to put up with my propensity. So, we put the goat in a pen and fed it and used it in class and whatnot. All of a sudden, we started having trouble with the data sheets. Something was eating them. These were the data sheets for the cow research that hung on the outside of pens where we were collecting data on clipboards. But what was eating them? Who knows? Every morning we would go in, and the goat would be right where it belonged. I started getting suspicious, so one time I looked in the window and rattled the door with a stick. Lo and behold, the goat was out. It tore down the alleyway and jumped into its pen. By the time I had the door open, the goat was right where it belonged. Well, this wouldn't do. Each number that's written on the data sheet is a pretty expensive thing, and to have a goat eat them is not

tenable. So, I took the goat to the farm. There are two things about the goat that are remarkable in terms of its experience at the farm. First, we have trouble with face flies. Face flies are a specific species of flies that gather around the face of cattle and other grazing animals, including goats. They are attracted by the elevated and triangular shape. That is, you can build a triangular thing and put it on a post and the face flies will congregate on that triangular thing. It's the shape and the elevation that attracts them. Once they get there, they feed on the nasal and eye excretions. They are an irritant to the animal and they also carry disease. Pink eye is the worst one that they carry. It's a problem. You normally control the flies with insecticides, but you don't like to keep spraying and treating with insecticides all of the time. Well, we were having a lot of trouble that summer with face flies. We had a cellar hole with round stones in the wall right near the corral. I noticed that the goat would jump down into that cellar hole and lean against the wall. I watched it and thought, what on earth is it doing that for? Then all of a sudden, I recognized what was going on. When the goat was leaning against the wall, its triangular silhouette was broken up and it did not attract face flies. While it was in the cellar hole leaning against the round stone wall, it didn't have the right triggers for the face flies, and they went elsewhere. As long as it was leaning against the wall, it was fly-free.

The goat stayed with the herd and accepted being a peripheral cow for the rest of the grazing season. Then, hunting season approached. The goat was brownish and not too different from a deer as far as general appearance, so I had to do something. I couldn't just leave it up there because someone was liable to shoot it. The afternoon before deer season opened, I decided to move the goat. The herd was up in the Owl Rock pasture, which is a long way from the road. The goat had a collar and it had horns. I went up and caught it—it was quite friendly—and I started leading it back to the truck, which was down on the road. Well, the goat got the idea that this was not going to be a good thing, so it refused to come. I pulled it along, but it would stop walking and drop flat on the ground, with all four feet sticking out one side. It did not want to go, and it figured that was the way to stop it. I ended up dragging that goat several hundred yards. The peculiar part of the story is, people were driving by scouting for deer season, and here's a guy dragging a brown creature looking something like a deer across a pasture the day before the season opens. Some people stopped and yelled, others honked, others stared and put their binoculars on me. It was an interesting situation. Anyhow, I finally got the goat down to the road and took it back to the lab. We penned it up so it couldn't eat the data sheets.

Bull Rolling Cow Over

This happened on a trip to the farm to feed the cows in February. I got there, and here is a cow on her back, in a little depression in the ground. She was quite pregnant and quite wide and very stable in that upside down position. She couldn't move. Now, a ruminant upside down is in real trouble because their rumen fermentation produces gas on a continuing basis and when they are upside down, they can't burp. The gas pressure just builds up in their rumen, and eventually it will immobilize the diaphragm and they will die. Simple as that. They are in real trouble. It takes about twelve hours for that to happen, depending on how recently they have eaten and the standard variations in the fermentation rate. Anyhow, the cow is in real trouble. Now, it's wintertime and there is lots of snow around. The tractor is in a snow bank up on the hill, so there's no possibility of getting the tractor to her. Also, the snow is so deep I couldn't get the pickup truck near her either. OK, Jim, what are you going to do? Gotta do something. What can you do? Well, I got a rope. I put it around

one leg of the cow—the leg was sticking straight up in the air—and I went around the other side of her and pulled on the rope and rocked her back and forth a little. She weighed probably 1000 or 1100 pounds and she was very stable. I could rock her a little, but I didn't have a prayer of rolling her over by that method. There wasn't any tree I could attach a block and tackle to, or anything. It was a mess. OK, Bright Eyes, what are you going to do here? Well, I'm struggling and feeling bad and thinking, gosh, I may lose this cow. And lo and behold, the bull comes along and starts watching me. He watches me very carefully. Then he goes around the other side of the cow and just as gentle and proficient and powerful as could be, he put his head and his horn underneath her and rolled her over. The cow lay there for a minute in a normal vertical position and, of course, she burped copiously. After a bit, she got up and burped some more. Then the bull just licked her face. Really just beautiful, just beautiful. I've told that story to a lot of people. Some people accept it. Some people try and dig a way out, you know, the bull was doing this, he really wasn't doing that. Or, this is really a phony story and you're just imagining the bull did that, he really was trying to fight with the cow and all this. Nah. None of the above. That's what happened. That's really what happened.

Horse

A student of mine came into my office one day and said, "Dr. Welch, I have a deal for you." She was half in jest because my reputation in the department was not as a horse-lover. I said, "What is it?" And she said, "I can get you a horse cheap." I said, "How cheap?" She said, "For $250 you can have a registered quarter horse, and you need one for your beef operation." I thought, maybe she's right. To her surprise, and a little bit to mine, I said "Well, let's investigate this further." She kind of blinked and said OK. So she arranged to see the horse. It was over in New York, west of Plattsburgh, and lo and behold, we went to see a horse. Number one, he was terribly thin. He was starving. That's why they were getting rid of him. He was out of grass. He had chewed the pasture down, and they didn't want to put any money in him and it was time, something had to be done. Well, the owner got out the horse, and put him on a lunge line and put him through his paces. It was a well-trained horse. The owner would shake the bridle and the horse would go into a show stance, legs all spread apart and his head way up in the air. And I thought to

myself, this is a nice horse. I also had some sympathy because they were going to put him on the meat truck. I said, "Sure. For two hundred fifty dollars, I'll take him." So, we got Horse, and he proceeded to eat the way a horse can eat. He ate twice as much as a cow. That's not documented, it just seemed that way. Of course, he had a lot of catching

up to do. And everything was fine. I got a saddle for him, which is still at the farm, you know. I would saddle him up and take him for a ride, which didn't work out too well. I still don't know what was wrong. The worst was when I took him over to the cows. If he was going to be a cow pony, he had to get into this thing. At first he was afraid of the cows, and he would run. This was with me on his back, which is a little less than some other things, me not being much a horseman anyway. After a while, the cows got a little more cautious and he got a little bolder. Then he started chasing the cows, which was worse. I tried several times to work through this, but I never had the knowledge base to make it happen. I'm just not that much of a horseman. So, time went on and he got more and more weight on him. Then fall came along. And, ah dear. He started implying to me that he needed a barn. I told him that he was a quarter horse, a registered quarter horse, and quarter horses in the West just turn their butts to the wind and get through winter that way. I told him that's what a quarter horse is all about. And he said, not this one. This one needs a barn. Jim, where is the barn? This argument went on for some time. He wouldn't go down in the draw to get out of the wind. I tried leading him down, and he'd look up in the trees. He just didn't like trees over his head. I tried to take him to places up in the woods to get out of the wind. No go. None whatsoever. He just would not do it. I'd stand there and argue with him. I'd say, "This is how it is, Horse. This is what we've got. Get used to this idea." Then he'd put his chin right on my shoulder and whimper in my ear, Jim, where's the pony barn? I never had trouble with him getting out or running away. But one day, I was driving through the gate out of the pasture. I look in the mirror, and he's got his chin over the tailgate on the pickup truck. Oh, god. Well, I thought, I'll get some grain and get him back in the pasture. But once he got out, he said, I'm out and boy, we're going to have a party! I worked around him some and struggled a bit and talked with him, but he wasn't going to go back in that pasture. Just was not going to do it. Then he started running. He went through the stream and up the other side and got in the road in front of the trailers. Oh, my god. What am I going to do? I thought, I can't let him get away. So I got in the truck and made the mistake of trying to get in front of him. Well, I found out later that he had some thoroughbred racehorse blood in him, that's what he'd been bred for, which is why he didn't take the quarter horse business too seriously. He was really a racehorse. We got going down the road there. I looked down at the speedometer, and we're going 35 miles per hour. He cuts into the pasture that's below the barn at the four corners there. Only, there's an electric fence around it. He hits the electric fence and gets it wrapped around his neck. It's still connected to the barn, so he's getting shocked. I thought he was going fast before, but now he's turned on the afterburners. There were a bunch of our neighbor's heifers down at the far end of the pasture by Duffy Hill Road. They look up and see this apparition running at them, and they went through that fence at the far end in five different places. They just obliterated it. It was a barbed wire fence, but they just obliterated it. They ran out in the road, ran up in the woods, ran out in the hay field down there. And Horse just kept on running. I'm sort of being a spectator at this point and wondering, what on earth is going to happen next? He's running and running down the road. I've given up trying to head him off. All I could do was just stay with him. And he runs and he runs and he runs. Down the road, near the B&B, there's a bend and a house right beside the road there. Well,

Horse ran up to that house and on to the porch. It was a cement porch with a green indoor/outdoor carpet on it. He stopped, and I thought, oh, thank goodness. But then the screen door flies open and four or five kids come out on this porch, screaming and yelling. I thought, oh, my god, what's going to happen next? Well, some place in the history of Horse there were kids. He just stood there while these kids pawed all over him. He put his head down and they scratched his ears. A real love fest. By this time, I'm standing in the driveway beside my pickup truck, just watching all of this. Then the mother shows up. She looks at the horse and she looks at me and kind of smiles and says, "Is this your horse?" And I say, "Yes, ma'am."

Then she kind of talks to it and hangs on to it. I get my belt off and I put it around Horse's neck because that's all I had. She says, "Wait a minute. I think we can give you a hand with this," and she produced a halter. Then she said, "I'll tell you what. I'll drive your pickup truck home"—she knew where Horse had come from—"and you walk him home and then you can drive me back." I said, "That's awfully nice, ma'am. Thank you." So with a halter on him, Horse and I walked home, probably a mile and a half. Quite a reasonable trek. Walking home, I come by the pasture where the heifers had gone through the fence. The owner was out fixing the fence and kind of shaking his head. I stopped and apologized and told him my horse did it and asked him if he wanted a horse, cheap. He looked at me and laughed and laughed and said, "No, thanks." Anyhow, I got Horse home and within a week, a guy stopped down by the road, climbed over the fence, and said, "That's a beautiful horse. Any time you want to sell that horse, here's my phone number." I didn't jump up and hug him right then. I was reasonably restrained. But the next day, I did call him and he came and got Horse. You know where he lives now, on Route 105. We see him often.

Young Bull

One of the problems with running a purebred operation is keeping all the matings controlled, that is, appropriate bulls with appropriate cows. If you're going to run a purebred operation, one of the expectations and hopes is that you will have young bulls for sale. To have young bulls for sale, you've got to have young bulls around. Young bulls can be sort of like teenagers, full of spice and vinegar. They get excited and charge around and make a lot of noise and generally can be difficult to manage. On the farm, my system was to keep the young bulls on the opposite end of the farm from the herd during breeding season. One day, I needed to bring the young bulls down to the corral for some veterinary treatment. Coming down through the woods, the bulls got excited. They normally would just follow the grain pail, and I would give them a little grain when they got to the corral. But coming through the woods, they got way ahead of me and started yelling and bellowing, and one of them ran through the fence. Pure exuberance. Now, the cattle communicate with themselves audibly over long distances, and this young bull that ran through the

fence knew where the cow herd was. He immediately set off through the woods in the direction of the cow herd. Well, the cow herd was intact and had a mature bull with it that probably weighed twice as much as the young bull heading toward it. I went over to the main herd just to see what was going to happen. The young bull was out of control, and there was nothing I could do. He advanced on the herd making a lot of noise as if to say, oh beautiful cows, here comes the answer to all of your dreams! He was making a big racket, lots of noise, lots of bellowing. My anticipation was the big bull would meet him head on at the fence and there would be some type of battle. But this did not occur. Instead, the mature bull picked a piece of high ground 30 or 40 feet from where the young bull was going to come up against the fence. Then the mature bull stood sideways, basically striking a Charles Atlas pose. As the little bull came out of the woods making all of this racket, he saw the big bull, and he immediately went quiet. He stood there and looked at the big bull, and the big bull looked at him, and the little bull looked again. Then, the little bull turned away and started grazing maple leaves as if to say, this is really what I came for. I drove the little bull back to where he belonged. His dream of being the big guy in the cow herd was obviously not going to happen, so he was happy to go with me.

Talking to Cows

This seems anthropomorphic, talking to cows. But just imagine this picture. You're standing out in the middle of the field working on a fence or something. You're all by yourself, nobody else around, and a car pulls up on the road, maybe 500 feet from where you are. Two men get out, and they start walking toward you. They don't wave, they don't smile, they don't say "Hi," they don't say anything, they just keep walking toward you. This is likely to engender some feeling of concern. What do most people do, if they want to talk to you about almost anything? When they get out of the car, they will wave, they will smile, they will say hello, they will do something to indicate that they are basically coming in peace. This is a very primitive sort of interpretation, but I think it is real. So, talking to cows. Number one, it works. What is it doing? It is setting the tone of the encounter. That is, cows are very sensitive to noise, very sensitive to tone of voice. There have been times when we've been working cattle and things are not going too well, and I get a little excited. Lorri will tell me that I need to back off and calm down because my angst is

spreading to the cows. So when you're going to do something, you need to set the tone of the encounter by talking to the cows first. This is not to say they understand what you are saying, but they do understand your tone of voice and they are very sensitive to that.

For a long time before I got a trailer, any time I had cattle to move, I had to hire someone with a trailer. I learned very quickly that the standard cow jockey's methodology for loading cattle on to a trailer or on to a truck simply did not work with Highland cattle. Did not work. The difference being that the jockeys were used to working with dairy cattle, and in a dairy operation you cannot tolerate free-thinking cows. It just

doesn't work. The cows have to go with the flow or else they get sold. The cow jockeys are used to that, so they have whips and flappers and things to frighten cows to go up the ramp or into a trailer. But I learned real quick that this didn't work with Highland cattle. You just can't push them. What I learned to do is tell these people, "Go sit in the cab of your truck and I'll load them, and then you can drive away." At first they didn't like that one bit. In one case the guy said, "I'm not going to put up with that" and I said, "OK. Bye," and he looked at me and left. But that's the only way you can work it with Highlands. You have to talk to them and make them think it really isn't such a bad idea. Use a little grain, a little patience. If you try to put them on by shoving them with a gate or something, they will say, I'm not going to do that. So you back off, talk about it some more, and you try again with a little more grain, a little more push, and eventually they'll go on.

As things progressed, I got a bigger pickup truck. I put a rack on it with no top that would hold one or two cows in the bed of the truck. Then I could load them and take them wherever they needed to go without having to hire cow jockeys, which was a reasonably expensive proposition. Well, one day I'm loading a little bull up, he's probably 500 pounds. I'm in the back of the truck with the little bull, having difficulty with him. He was not happy with what was

going on, but I got him on board. As I was tying him up, he got his head under my backside and threw me right up over the top of the truck. I was backed up to a bank that's across the road from the corral, and I came down head first, which for me in particular is not a good exercise to go through. Fortunately, I hit on the bank at the appropriate angle,

and I bent and slid down the bank. So, here I am on my back, totally shaken. I mean, totally shaken. And I'm wondering, Jim, did you break your neck? I feel lots of pain, but I can't identify anything as far as the central nervous system. I start wiggling my fingers, and they all worked. I thought, try your toes. I wiggled my toes, and they all worked. After a while, I sat up and kept poking and trying this and trying that and everything worked. I got back in the truck and tied up the little bull, and we went on our way. That was a reason for buying a stock trailer. It was not a good thing to happen, and if I continued with that system, it was likely to happen again in some form or another. So, we bought an aluminum stock trailer.

Little Zach

This is what appeared at first to be a most unfortunate story. I don't know how much to beat on this, but the initiation of a calf's life is a very tenuous and hazard-filled process. They start off getting dumped on the ground with no Dr. Spock and nobody else around to help. They are expected to, and in most cases do, get up, find the cow's udder, get on a teat and nurse, often within an hour of being born. When that goes right, it is wonderful. But sometimes it doesn't, for various reasons. Anyhow, Little Zach didn't make it through that process. It was apparent that he just wasn't going to live if we left him with his mother. So, Lorri and I brought him to our house in Colchester. Now, in order to get the microbial situation balanced properly, a calf needs to get colostrum. Little Zach got some colostrum and then we started feeding him milk replacement, a commercially prepared milk-based product that has everything a calf needs in it. It is commonly used for raising dairy calves because it is cheaper than feeding them whole milk, which is a saleable product. Anyhow, we brought Little Zach down to Colchester and commenced to feed him milk replacer.

He started off all right. He had some other veterinary problems, which we got cured. So he was doing all right, but then he started going downhill. There was something wrong with his rumen microbial mix and the esophageal groove. In a normal nursing process, there is a groove that closes and puts milk directly into the true stomach of the calf. It bypasses the rumen and the reticulum, which is where the future grass fermentation is going to occur. Well, anyhow, Little Zach grew for several weeks but then he started losing weight. His esophageal groove stopped working properly and he started putting milk in his rumen, where the microbes worked on it and kind of made curdy cheese out of it, which was no help at all.

He started regurgitating this cheese-like product on to the ground, and he started losing weight and he stopped drinking. I don't know if this is kind of grim, but I had started to think we might have to put him down. He was struggling, he was miserable, and it's not fair to let an animal die in misery of a situation that he is not going to recover from. Well, Lorri went out and took some milk, some regular milk, and she washed him all over. She washed his little butt. She washed his face. She talked to him and she washed him some more and then offered him some regular milk and paid more attention to him like that. And just all of a sudden, he kicked in and he drank milk out of the pail and he took off. I hadn't given up all hope yet, but it was that close. So at this point he's two months old and rather stunted. He hadn't grown much at all, really. He'd grown a little, but then he'd gone backwards. Because of Lorri's TLC and concerted input, Little Zach started to grow and get on to a healthy life course, but he was still stunted. We nursed him along and he got so he'd eat a little grain, then eat a little grass and then things got better and better. Now, the problem arises. It looks like he's going to live, so what are we going to do with him? That is, he wasn't vigorous enough to go back with the herd on the farm. Just no way. And he was a pet, a real pet. So, what are we going to do? Well, I was talking to the feed man about what to do. And he said, "Oh, somebody over at Mazza's Farm is

starting a petting zoo and you might go talk to them. They might like a little calf." So, I go talk to them. The first time—you know Mazza's is a huge greenhouse outfit among other things—they were planting seeds and the guy was far too busy to talk to anybody because they had a whole crew

hammering away at getting seeds put in for the greenhouse business. So, I try again and finally we connect up. I tell him I don't know what's going to happen to this animal. He may never be a good, mature, healthy animal, but he's sort of on his way. Well, lo and behold, the man says he'd like to take Zach on, and we established a minimum dollar exchange for him. The guy over there at Mazza's just fell in love with him. And, Zach returned it. As you may know, Zach is still there. He weighs close to a ton. He has a full set of horns and he is one of the stars of the petting zoo. Little Zach. He is loved and well cared for. From being snatched from death's door to a very nice future for a steer. That's pretty good stuff. The best you could ever hope for.

Curiosity

This bull was Jupiter. At one point, I was on crutches. When we went to the hay shed for the winter feeding procedures, Lorri fed out a ton of hay while I walked around on my crutches and talked to the cows. One time, I came around the corner of the feeding shed and here was the bull, right smack in front of me. While Highland bulls are generally very tractable and not dangerous, this was a peculiar situation. I had one foot hurt and was using the crutches vigorously in the snow and ice and manure. The bull walked toward me, and I wasn't sure what was going to happen. An injured animal is more likely to be attacked, and I was clearly injured. I was not acting normal. Jupiter walked straight up to me. He had a large set of horns, probably four feet across tip to tip. First, he smelled the crutches very carefully. Then standing right smack in front of me, he took a horn tip and went up one crutch—tap, tap, tap—tapping every two to three inches up the crutch, and did the same procedure up the other side. I'm standing there thinking this is not a good situation. But after he finished tapping, he smelled them again and walked away. And my heart rate declined appreciably.

Willow

Willow's story starts with a successful birth. Her mother was Nymph, one of the most difficult cows we ever had on the farm. Whenever we rounded up the herd to do some sort of treatment, Nymph would run up to the top of the hill and look at us through the bushes. She wouldn't go near the corral. She was a difficult cow. She just was contraire. Nymph calved successfully, but then I went in to the hospital for something. The day I got out, I wanted to go to the farm to check on the cows. I made a deal with Lorri that I would just look—not drive, not do anything—so she drove me there in our Buick sedan. When we got there, we saw Nymph's calf had somehow gotten through the electric fence and was unable to get back. She was starving and in very poor shape. When we got them back together, she was too weak to nurse. I don't know what had happened there. That's one of those times when things come unglued and you don't know why. Most of the time it works beautifully. Out of 600+ calvings, we only have ten stories. It had possibly just been too long, but whatever went on while I was in the hospital, it tore things apart. Nymph being a very difficult cow, it wasn't an

option to hold her. The only other option was putting her in the shoot, dragging the calf up to her, and forcing the nursing process on her. There's one story of the cow that allowed me to teach the calf in pasture to nurse, but normally if the calf is not nursing, you have to take them to the corral because it won't happen by itself. If a normal process was going to take place, it would have already happened. The other thing, with a new calf, Nymph was dangerous and aggressive. So, what on earth are we going to do? We circled around in the pasture in the Buick. Eventually, we got between Nymph and the calf. Lorri reached out, grabbed the calf, and pulled it into the car. We were in the Buick, not the truck, so we had no ropes, no grain pails, nothing. The only solution was for me to drive and Lorri to be in the back seat with the calf. The calf was jumping all over the back seat, and Lorri was wrestling with it. It was a struggle. We didn't have it tied up because I had assumed it was so weak and starved that it would just lay there. Here we are, driving down the road in the Buick with a calf in the back seat. I look in the rearview mirror and see a motorcycle come roaring up behind us. Just at that moment, the calf gives a lurch and its full face can be seen in the window. The guy almost falls off his motorcycle when he sees the calf in the car. Well, Willow gets down to Colchester and we feed her and get her some colostrum and milk replacer and she prospers. She becomes a very large, healthy creature. Because of her pet status, when she got big enough, we sold her to some people with a small farm nearby who had fallen in love with Highland cattle. They kept her for more than a year, but then they dispersed their herd and we bought her back. We put her in with the rest of the herd on the farm. She was big, but because she was not raised with the herd, her social status was low. She was allowed to eat with the rest of the herd—if she had been really low down in status, they wouldn't have let her near the feed trough—so she could stay with the herd, but they ganged up on her and pushed her around. When we had the herd dispersal, we kept Willow. There were a couple of reasons. Number one, she was Lorri's pet.

Number two, she was not a successful reproductive cow. She was an intermittent breeder. She'd have a calf and then have a calf the following fall and then skip the whole next fall. I couldn't sell a cow like that. We had her down at Covenant Hills church camp two summers when I did an animal science unit there. I would bring some livestock and talk to the kids about it. She did very well. When we came to bring her home, she bugled and ran and showed obvious joy at seeing the trailer to bring her back to the farm. She required no herding. We just opened the door of the trailer and she ran on. She wanted to go home.

When the herd was dispersed, we sold all but five animals. We kept three females: Beauty, a small young heifer; Thistle, the old dominant cow; and Willow. Thistle hammered Willow unmercifully. This went on for a year and a half. Then one day I went to feed them, and who was in the front row, but Willow. Thistle was in the back. Thistle tried to approach the feeder, and Willow pushed her back. They'd apparently had a set-to, and Willow had asserted her physical dominance and ascended to the top of the heap. She is still a poor quality breeder. She had a calf in September, which is a terrible time in our breeding system. She is still lactating, on hay. She has huge body stores, which is where she is lactating from. It's not coming from the hay.

Grazing Trees

There were times when there was drought and the pastures got short. Highlands have a particular predisposition for browsing, that is, they like to eat tree leaves, more than most other breeds. Given saplings, they particularly like elm and poplar. They'll walk up to a poplar sapling and hook their horn on it. Then they'll walk along, holding the sapling with their horn, and chew the leaves off of it. This is one of the things that the Highland breed is touted to be good at. If you have brushy pasture, they will help you clear it. They won't eat everything. They don't like black alder and white birch, but they love maple, elm and other species. In a drought year, when you're really short of feed, the herd is hungry. What are you going to do? Get the chainsaw and cut down big trees. Big poplar trees, is what I did mostly. I would get the leaves down close to the ground where they could get at them. The herd would eat the leaves off four or five good-sized trees in a day. This would solve your feed problem. If they are really starved, they will eat the bark, too, but that's only when they are really starved. About the third day you do this, they figure it out. They hear the chainsaw and, ah-hah, dinner is served! You are standing in a grove of trees with a whole herd watching you. You want to say, herd could you move out of the way? Instead, you have to develop a strategy for moving the herd away from the trees until they are all down. You get a little grain, take them to another pasture, and close the gate behind them.

Of the time we had the cattle, I had to cut down trees to feed them three or four years. I knew they'd go for it. When you turned them into a new pasture, they'd hustle around and eat all of the saplings that they could bend.

The Cow Whisperer

This farmer I knew, a Highland breeder, had a capability with cattle that was most unusual. That is, when you went to his farm, you'd go out in the pasture, and he'd go up to a cow and talk to her. He'd stand there and put his elbow on her tail-head, you know, leaning on the cow. Most of his cows were friendly like that. He could tame them. He would take a cow and a young calf to fairs and not be worried about somebody getting hurt if some little kid went up to the calf and made it bawl or something. He just had an incredible way of taming cattle, calming cattle, and getting along with cows. I don't know what he did. I watched him time and time again. I know he didn't lift his feet. He didn't prance, he shuffled. He never showed them the inside of his hand, always the back. And he talked in a very low voice, which I suspect was a major part of it. I'll just tell you one story of the kind of thing he could do. Just amazing.

He wanted to buy a yearling bull from me. There were five yearling bulls up in the pasture through the woods over by Route 108. He brought a big stock trailer with him and left it by the road, down by

the corral. We walked up through the woods to the pasture, and he picked the bull he wanted. I had a grain pail, and I would have led all five down to the corral, cut out the one that he wanted and returned the other four back to the pasture by having them follow the grain pail. I described that to him and he said, "Oh, we don't need to do that. He'll come with me." I looked at him and said, "OK, I'm watching, I'm listening." He walked out in the pasture and walked up to this bull, and stood there talking with him. A few minutes later, he's rubbing his neck and talking in his ear. The bull is just standing there looking at him, taking in all this attention. I could no more have done that than jump over the moon. I don't remember exactly what he said, but basically, "Come on, let's go," and the bull followed him. The other four bulls just stood there watching. I opened the gate and he and the bull came through. The bull followed him all the way through the woods and down to the corral. I got down there and scrambled around the back of the trailer to open the big back door. He said, "No, don't do that." There was a little side door that was much higher than the back. He said, "He'll come in here with me." I thought I had heard wrong. Never in the world would that bull jump up like that to get into a strange trailer. But the man stood there in the elevated doorway and pretty quick, the bull jumps in. Just amazing. Absolutely amazing.

Herd Dispersal

The herd dispersal was a traumatic time for all concerned, as it turns out. We averaged about 60 head during the grazing season, and it would trim down to 45 head over the winter. When we decided we couldn't take care of our 60 head appropriately any more—the work just wasn't getting done—it was time for herd dispersal. We were very fortunate in that during the herd dispersal process, almost all of the animals went to good breeding herds. There wasn't the trauma of having to ship them all off to the slaughterhouse. That didn't happen. Some went to better placements than others, but they all had a good chance. I knew that herd dispersal was traumatic for the owners, but I didn't understand how traumatic it would be for the cattle. There were two things that specifically happened. Number one, after the first couple of trailer loads came, loaded up, and left, we had major difficulty getting the rest of the herd into the corral if there was a trailer around. They just would not go into the corral with the trailer there. Animals would get loaded on and they never came back, never to be seen again. The herd didn't know that they went to

good homes and all that, upward and onward. They just knew they disappeared and that wasn't good. The bulls in particular wouldn't go near the corral at all. They'd stay down in the alders by the stream and kind of peek out from behind the bushes and watch what was going on. If there was a trailer there, they wouldn't go near it. You could wave grain in their face and they wouldn't follow it, as if to say, I'm not buying that. So, we had major difficulty particularly loading out the bulls. They were not compliant, and they apparently understood some of the gravity of the situation. The other difficulty was Thistle. One of the cows I chose to keep was Thistle. She was a very productive cow, a very nice cow. I enjoyed her. She was one of my favorites. But after the

herd got down to five, she would try to kick me every chance she got. No apparent reason whatsoever. In the corral, out in the pasture, any time I got near her she would try to kick me. There was no indication of that kind of behavior before. I chose to keep her because I thought she was such a nice cow. But after the rest of the herd left, she was cross at me, at least that's what it looked like. I learned to stay out of her way and not get in a position where she could nail me. I wasn't doing anything to prompt her. I wasn't pushing her from behind or trying to get her to do something. I was just walking around the corral with the grain in the pail. If I walked anywhere near her, she would try and kick me. After it happened a couple of times, I was very wary of her. And before that, there was no indication of any behavior pattern like that. You can call it what you will, but that's what happened. There were no other effects on the other cattle I ended up keeping. I kept Thistle for five years. I kept her around because she was a very productive cow, but I sold her last year. Her calf is still here. Her name is Beauty. She has a calf every year in May and does very well. We're eating one of her calves right now.

As an aside, out of the 100 or so animals at the 2009 National Western Stock Show in Denver, there were eight progeny from our farm. We talked to the people who brought them, and they like them. It felt good.

About the Authors

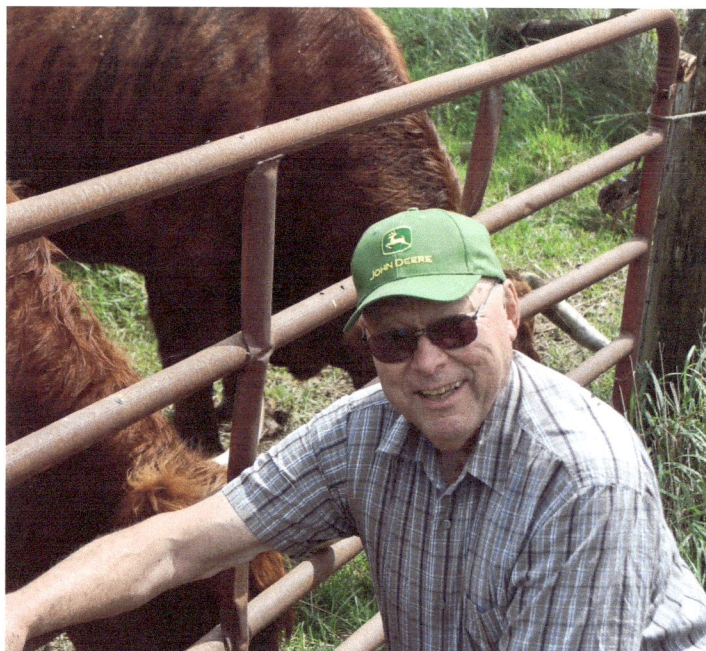

Dr. James G. Welch is Professor Emeritus of Animal Science at the University of Vermont. He raised Highland cattle on Owl Rock Farm in Franklin County, Vermont for almost four decades. He served on the American Highland Cattle Foundation Board of Directors for many years. He lives in Vermont with his wife, Lorri.

Dr. Anne S. Welch grew up listening to her dad's cow stories. She enjoyed working with him to write the stories down so they could be shared with a wider audience.

www.ingramcontent.com/pod-product-compliance
Lightning Source LLC
Chambersburg PA
CBHW041652260326

41914CB00017B/1618